Psychic Matter & Strange Particles

The Contemporary Handbook of Science & ESP

PRISCILLA A. KERESEY
SHAE MONTGOMERY

LIVE & LEARN
New York

Copyright ©2018 Priscilla A. Keresey & Shae Montgomery

Editor: Peggy Bartlett
Cover Design: Shae Montgomery
Portrait Photography for Priscilla: ©Lisa Marie Pane

All rights reserved. No part of this book may be reproduced or transmitted in any form or by any means, electronic or mechanical, including photo-copying, recording, or by any information storage and retrieval system, without written permission from the author, except for the inclusion of brief quotations in a review.

Published By
Live & Learn
P.O. Box 226
Putnam Valley, NY 10579

Orders: www.liveandlearnguides.com

ISBN: 978-0-9863536-3-5

Printed in the United States of America

"Science without religion is lame. Religion without science is blind."

Albert Einstein

TABLE OF CONTENTS

Introduction	9
Chapter 1: What To Expect	11
Chapter 2: Then & Now: What Are We Talking About?	13
Chapter 3: Getting Your Metaphysical Mind Ready	21
Chapter 4: Get Started—Who & What Am I	28
Chapter 5: Mind & Body Pushups	32
Chapter 6: How Does This Fit in My Daily Life?	54
Chapter 7: The Ethics of ESP	59
About the Authors	69
Also by Priscilla Keresey	71

INTRODUCTION

What you're about to read is evidence that scientists and psychics *do* have common ground, in fact a great deal of it. The authors of this narrative come from profoundly different occupations, beliefs and experiences, yet discovered that the bridge between them has shared foundations in science, education, historical thought, and current theories of the state of humankind and the world(s) we inhabit.

In this book you'll see for yourself why extrasensory perception (ESP) is not just a party trick, and hear how scientific exploration has been trending in the direction of "proving" the sixth sense for decades, culminating in present day.

The definition of metaphysics is "a division of philosophy that is concerned with the fundamental nature of reality and being" and "the study of what is outside objective experience." We know from both the scientific and "New Age" perspectives that the subject of metaphysics is exciting, not yet fully defined, and therefore – if you're open-minded – an area ripe for ongoing exploration and personal advancement. Metaphysics embraces theories of the Multiverse, telepathy, life after physical death, and contemporary ideas about the human brain's ability to grow, adapt, and change.

Being rational while staying tuned into the non-rational is a balanced (and fun!) way to go through life. We all know that person who is so "imaginative" that he or she never seems quite grounded, and floats through life absent-mindedly or deflecting responsibility. We also probably all know that person who is so super-logical that he or she has no room for imaginative or spontaneous interaction. A healthy goal is to find the balance between these two extremes. By trusting both your logical mind *and* your natural, built-in extrasensory perception, you'll gain confidence, feel more decisive, and be better able to anticipate roadblocks and successes. That's because the sum of these combined input channels is so much greater than the total of the parts.

The authors invite you to dive right into this book, engaging both your thinking, rational mind, and your unlimited and staggeringly powerful imagination. Try out the exercises; there are *plenty!* Some may resonate with you more than others. It's not necessary to do every single one to gain that special and empowering discernment that is the product of metaphysical study.

We make no guarantees except that your thinking is likely to change profoundly, not just about your own extraordinary abilities, but about the world and your place in it. Be prepared for constant validation that your brain and your mind are changeable in both proven *and* unforeseen ways, and that your own natural sixth sense has more information to share than you ever thought possible.

We'd love to hear about your experiences, so please stay in touch with us. You'll find our contact information at the back of the book. We continue to explore our own abilities and have many ongoing ideas we're manifesting, including the power of intention, new protocols for global healing, living thoughtfully in the multiverse, and experiments outside of our regular, three-dimensional world.

Enjoy!

Priscilla Keresey
Shae Montgomery
September 2017

CHAPTER ONE
What To Expect

Extrasensory perception, or ESP, has been known by many other names: women's intuition, fortune-telling, insight, clairvoyance, second sight, hunches, gut feelings, divining, and sixth sense, to name just a few. In this book, as with all the books written in this collaboration, you'll learn the current science behind ESP and how to access and take advantage of your own natural abilities. You'll learn the role of brain function in ESP, how to identify what intuition feels like, and how to read the signals from your sixth sense.

Just as there are many ways to talk about ESP, there are equally as many ways to experience ESP: by clairvoyance (or clairsentience, -audience, -gustiance, -cognizance, and -scent), psychometry, divining tools such as tarot cards, runes, tea leaves, palm reading, and crystal balls, telepathy, precognition, retrocognition, and mediumship. These all have ties to current thinking in physics and medicine as well as links to your brain capabilities. We outline the science aspect in a digestible fashion throughout this book as we talk about experiencing ESP.

In the following chapters, you'll learn the extraordinary benefits of accessing and using sixth-sense information in your life. You will gain an understanding of:

- Why the connection is possible
- Society's current and historic use of ESP
- ESP basics and how to start using your new knowledge
- Incorporating awareness and practice into your daily life
- How to notice and utilize the changes that occur
- Ways to harness and continue to improve your ESP skills
- The ethical use of ESP
- How to handle existing social resistance
- How physics and metaphysics relate to ESP

You may notice that we, the authors, come from two different fields of study: science and metaphysics. Because neither one of our perspectives can be proven with any certainty, we have deliberately sought out and celebrated our diverse views, discovering common ground in the process.

Spiritualists and mediums use their sixth sense to connect with the living energies of beloved friends and family; more esoteric practitioners use their sixth sense to examine past lives, channel Divine voices, or connect to spirit guides.

Because our rational mind awareness (the scientist in us) and our sixth sense awareness (the believer in us) are always evolving, no one can say with absolute certainty that a static state of "truth" exists in either. Maybe you're on the fence about what and whether to believe. We think that's exactly the right place to be!

Uncertainty coupled with openness offers the greatest personal development no matter what you are studying.

We believe we're both on to something, the scientist and the psychic. The common ground where we meet is the sixth sense. As pioneers in this as-yet-unmapped frontier, we're working together, supporting and learning from each another, even if our beliefs don't exactly match. So remember as you develop your own awareness of your ESP that there is no right or wrong, and the science behind ESP might just be the grounding you need to go forward. Or perhaps the mystical behind the science provides that lift you need to move ahead.

Have fun! Be relaxed about your understanding as it develops, recognizing that even within your own process, your beliefs are bound to evolve. Whenever you hear a psychic, scientist, or any "guru" rigidly declare "the truth," take it with a grain of salt.

CHAPTER TWO
Then & Now: What Are We Talking About?

Do You Ever Feel Like A Neanderthal?
Did you know that there's a little bit of Neanderthal in many of us? According to recent research, apparently some Homo Sapiens (that's us) crossbred with Homo Neanderthalensis (Neanderthals). Interestingly enough, our cousins the Neanderthals may have had more of what we might call "data storage" in their brains, but they had smaller forebrains. The forebrain is actually what differentiates us from animals.

It's important to briefly explain the anatomy of the brain, as it is intimately tied to communication and ESP. There are three main regions of the brain—Hindbrain, Midbrain and Forebrain. As mentioned, the forebrain is a key differentiator for humans. The forebrain contains the cerebral cortex, which is the most recently developed part of the brain (according to studies done by humans*). Why is the cerebral cortex important for ESP? It contains regions of the brain that control movement, planning, reasoning, personality, language, hearing, vision and overall perception, long term memory, and general thought processing—all elements involved in communication, and thus ESP. So for our purposes it is a good thing Homo Sapiens, with their larger forebrains, won out!

> (*By our own definition of bias, any experiments that we are a part of are flawed. Any experiment that we design is fundamentally biased because we are the designer.)

Follow Your Gut!
ESP has been proven and disproven over the years, depending upon the source issuing the evidence. Joseph Dunninger, a notorious mentalist, noted this each time he signed off his broadcast TV show

when he famously quipped, "For those who believe, no explanation is necessary; for those who do not believe, no explanation will suffice." Keep this quote in mind as you begin your own journey.

This book is not intended to convince anyone or prove that ESP is real. We assume that if you're reading it, you are at least open to the possibility that extrasensory information is available to you. No matter what you call it, every person alive has a sixth sense. Every person has experienced it, whether we're talking about a mother "just knowing" her child is in trouble, a police detective following a hunch, or a salesperson identifying who is truly interested in buying. The gut feeling is built into our evolution, and we pass it along in our DNA. We needed it to survive as a species and it continues to evolve to ensure our species' survival.

While we're not attempting to convince you, we do want to give you the information that is out there, both to build your confidence in your own skills to make you more effective, and to provide you with information to pass on to those in your care. We'll cite many examples of psychic ability in this book, but let's start here with one from the government side of things. The U.S. Federal government spent 20 years and 20 million dollars investigating psychic phenomena, and how it could be applied domestically and militarily. This project, called the "Stargate Project," took place from the 1970s through 1995, and was primarily handled by the CIA and DIA. The project started with research from Stanford Research Institute in conjunction with the Science Applications International Corporation. The Federal government determined in the end that "even though a statistically significant effect has been observed in the laboratory, it remains unclear whether the existence of paranormal phenomenon, remote viewing, has been demonstrated." And the government further concluded, "even if it could be demonstrated unequivocally that a paranormal phenomenon occurs under the conditions present in the laboratory paradigm, these conditions have limited applicability and utility for intelligence gathering operations." After 20 years and 20 million dollars the public facing statement was that we found some evidence but we aren't entirely sure how to use it. I'll leave you to digest that one for now…

Sixth Sense? I Can't Even Use My Five Senses Properly!

One of the fundamental tenets of ESP—no matter what you call it or how you apply it—requires understanding and acceptance that information from the sixth sense does not feel the same way as information that comes through your other five senses, or your "rational mind". Often, ESP doesn't even make much sense at the time because as a society we don't yet have the language to talk about it comprehensively. We're sure you can cite several examples of déjà vu from your own experience, times when you watched an event unfold and said to yourself, "I knew that was going to happen!"

We believe that there is nothing weird or mystical about your sixth sense. It's simply another channel of information that you can use to help make good decisions. You might be surprised to learn that your brain chemistry and even the molecular makeup of the air around you affects the availability of those information channels.

Learn How To Save The Day!

Just as your vision, hearing, and sense of smell might offer evidence of fire, your sixth sense can also offer evidence of a future event like a fire (called precognition or premonition) or evidence that a fire had taken place (retrocognition). We have a neighbor who always had a "funny feeling" when he stood in his living room that the house he bought had had a fire, though the real estate agent assured him it wasn't true and the structure was sound. Years later while putting down new flooring, he discovered charred beams that had been covered up by the first owner. At the time, our neighbor hadn't smelled or seen any evidence of the charred beams, as the fire had taken place more than thirty years prior. Yet, on one level a part of his mind was letting him know that it had indeed happened.

He remembers people laughing at him and rolling their eyes when he talked about his feeling. It made no rational sense, he couldn't prove it, yet the notion of a fire in his new home persisted and ultimately bore out.

We present this example for four reasons. First, to illustrate that psychic impressions seldom make rational sense at the time; second, that people will often ridicule or undervalue what cannot be "proven," third, psychic impressions persist, and fourth, we have many brain functions and abilities that we have yet to understand and we continue to underestimate the senses that we already acknowledge.

When you experience an episode of ESP, you will also notice a few other qualities that differentiate those impressions from regular conscious thought. We'll touch on these with specific examples in the Get Started section. For now, rest assured that you do have a sixth sense that you are capable of accessing and understanding, and there is science behind it. There is no need to be gifted, religious, spiritually special, or born under a certain star in order to make good use of all the valuable information available through ESP.

How Do I Drive This Baby?

To further expand on the science behind ESP, let's talk about anatomy and quantum physics. On the anatomy side, we constantly affect and change the neural pathways (or channels) in our brains. These pathways are governed partially by chemical activity (neurotransmitters) that we can directly influence by changing the way we think and act. Our thoughts create, solidify, and change neural connections, and emotion is tied to this process as well. For example, if a memory is rarely stimulated (thought about) or has no emotional attachment, eventually the neural connections associated with it will diminish from disuse. We can actually use this knowledge to make our brains work for us, and finely tune our ESP skills.

On the quantum physics side, with the recent news confirming inflation (the theory of exponential expansion of space in the early universe), we are finding more and more plausible explanations for "phenomena" like psychokinesis, ESP, and paranatural occurrences. Most models of inflation include the existence of the Multiverse (the concurrent existence of many universes at once), which opens up entire cans of worms for continued exploration by our scientific community. The Multiverse includes the phenomenon of entanglement, which helps support ESP, altered consciousness, the possibility of consciousness without a physical brain (implying life after death), synchronicity (meaningful coincidence), and more. We'll talk more about the Multiverse in the next chapter of this book.

Once we understand how our brains work and the role science plays in ESP, we can refine our brain patterns and our world to hone this skill. Read on!

Why Hasn't Anyone Thought Of This Before?

Well, they have! In fact, throughout history experiments have been done to test ESP. At different times in history, science, philosophy, religion, and medicine were considered one thing. Therefore, ESP experiments included all of these disciplines. The famous philosopher and scientist Aristotle (384-322 BC) was fascinated by dream telepathy (telepathy being under the larger umbrella of ESP). He argued that telepathy was possible in a non-dream state as well. In one essay, he compared the ripple effect of a stone thrown in water to a telepathic message. He wrote that the mind propagates waves just as waves are propagated by the stone. Aristotle together with Democritus made telepathic dreaming part of scientific inquiry; science and philosophy together.

Another important scientist, Alfred Russel Wallace, believed in contacting other souls through spirit mediums. How influential was he? Well, his independently-produced theory of natural selection was published in part with some of Charles Darwin's writings in 1858.

In 1914, Sir Arthur Conan Doyle (physician and writer) gave up writing the popular Sherlock Holmes series to investigate psychic research as one of the leaders of the Society for Psychical Research.

And our dearest Freud! In 1933 Freud wrote in an essay that he believed in telepathy and saw psychoanalysis as a way to explain phenomenon in the physical world. He went on to say that telepathy may be a remnant of an earlier method of communication that had now been replaced by speech.

But Science Can't Prove ESP

There was a time far back in human history when unexplainable events were given mystical status simply because we didn't know how else to be comfortable with things we really didn't understand. Superstitions, gods, and rituals arose as a way of making sense of thunderstorms, fire, and other natural weather occurrences. Once humans began to document and culturally incorporate the rules of the universe according to their understanding, contradiction of these rules usually resulted in expulsion from the church, university, or community. Cast your mind back many centuries, when the scholars of the day were certain the Sun revolved around the Earth. Opposing viewpoints were immediately squashed under threat of torture and death.

We now know a lot more about weather, planets, physics, and brain functions than we ever did in the past. But that doesn't mean we're finished! Learning more about the complex organisms we are settles some questions, but raises many more, including the currently unexplainable phenomena of ESP. There's no way yet to reliably measure ESP, so for those who demand—and won't get—a "rational, scientific explanation," it's confirmation that ESP simply doesn't exist. But just because there's no way to measure it right now, doesn't mean all exploration should cease.

When Wallace, Doyle, and Freud were forming their ideas about the sixth sense, you can bet there was quite a bit of resistance. That's because there are Doubting Thomases around the leading edge of every "new age" idea, trying to dismantle progress. So, don't be surprised if the experiments you try in this book make other people laugh, or dismiss them altogether. What we don't understand creates curiosity in some, and fear in others. That fear can look like derision, mocking, and wholesale dismissal of new ideas. Don't let it stop you! Nothing feels more satisfying than validation. Remember our neighbor with the fire in his house? Even though fixing the decades-old damage cost him a small fortune, he says it was all worthwhile when he discovered he wasn't "crazy."

You Probably Already Do This

There's a fairly recent approach in veterinary care nowadays. Our companion animals can't articulate exactly how they're feeling when something's wrong, so they rely on a human member of the family to communicate for them. Most good veterinarians will seek out your opinion because they know that you know the animal best. We brought our German Shepherd Memphis into the vet when we noticed he was acting slightly clingier. I simply said to the vet, "Something's wrong." She sat down with me and asked me why I felt that, and what my feelings were about a possible cause. By giving me the chance to share my feelings about my dog's slightly "off" behavior, the vet was able to create a treatment protocol. Turns out my gut feeling about Lyme Disease was correct, and Memphis recovered beautifully after a course of antibiotics.

Memphis wasn't able to tell me that he had Lyme spirokytes racing around in his blood stream. He wasn't limping, lethargic or displaying any other extreme signs. Yet I had the hunch (otherwise known as ESP) that something was wrong. I was sensitive to it because I love Memphis and am tuned into him even though he's an entirely separate being from me. Imagine how you can tune into your own best options if you know how to read the signs!

During our classes on ESP skills, many of our students express fear that they're only going to see "bad" things. If you practice the exercises in this book, you will never see bad things. Using your sixth sense empowers you to increase your good judgment. Developing your personal ESP enables you to discern more clearly just what is (or isn't) going on around you in your health, love life, finances, career, and family.

Chapters 4 through 6 contain examples and exercises that show how to tune into your own ESP. Remember, you don't need to create your sixth sense. You already have one! Rather, you need to harness and develop it. In the same way that you presumably were born with vision, hearing, and the senses of smell, taste, and touch, you were born with a sixth sense. If you were born lacking one of these senses, or lost it at some point during your life, your awareness of your sixth sense may even be stronger than average.

Notice we said "awareness" of your sixth sense. You have a vibrant, usable, perfectly functioning ESP, whether you take advantage of its information or not. Your brain is set up for it. You can't escape it! It's a necessary part of your very survival. So you don't need to have a special guru open a chakra, touch your third eye, or cast a spell on you. You don't need to practice for years in a cave or go into a deep hypnotic trance. You simply need to follow the easy, fun exercises in this book and learn how your sixth sense is talking to you.

And if those Doubting Thomases interfere—or the inner skeptic in your own mind interferes—refer to the science in these pages. We might not be able to prove that psychic phenomena is "real" in the scientific sense, but we can show you that your brain is wired to process non-rational information as well as the information from your other five senses.

CHAPTER THREE
Getting Your Metaphysical Mind Ready

A Person Who Makes A Mistake On An Elevator Is Wrong On Many Levels

It is important to understand what we do know as a society currently, and what we do not know. And it is also important to understand that we are all still learning, that science is changing everyday, and that science is often proven wrong. These two statements may seem to conflict, but the message is that knowledge is power as long as you are willing to be wrong. Wait, that is in conflict too! Sufficiently confused? Good, now your mind is open and you're ready to get started!

Let's Get In The Know

As mentioned previously, ESP can be approached through quantum physics as well as anatomy. Let's first try to understand the basics of quantum physics that that support this reasoning. To greatly simplify, quantum physics concerns itself with energy, matter, subatomic particles, physical systems, and their relationships. The existence of the multiverse and the theory of entanglement are two current discussions in quantum physics that help explain the existence and practice of ESP.

The Multiverse posits the existence of many universes thriving all at once, possibly an infinite number. This theory is still being debated, but it's gaining more and more acceptance by prominent physicists, especially with the confirmation of a phenomenon called Inflation.* Many physicists agree that it is unlikely that we exist in a single universe, even if they don't think the multiverse theory is correct. Many theories have been proven and disproven in the course of science. At one time, the roundness of the earth was a theory. Then we accepted that the Earth was round, but we thought that the sun revolved around it. When we accepted that the Earth did indeed revolve around the Sun, we still thought it was the only planet. Then we accepted that the "wandering stars" were other planets. We now know that other solar systems, and galaxies, and other universes exist.

In the Multiverse, time, communication, and relationships function very differently than we understand them in this universe alone, because they are all altered and affected by their parallel existence in other universes. Many copies of our theoretical selves can be out there living many different lives in many universes that have different physical properties than ours. And these multiple copies of ourselves in many universes can help to explain intuition and mediumship, as well as empowering you to develop the necessary skills to access them.

How could we not obtain information from and be affected by our other selves? At the very least, if there are no other "selves," inhabitants living in parallel universes and events occurring there would affect us. When people exit this universe (the deceased), but are likely to continue to exist side by side us in a parallel universe, why would we be unable to communicate with them? If there are multiple or infinite parallel universes for which time exists differently, how could your loved one not exist in one of them… after all you are.

Quantum Entanglement also helps explain the infinite connections we access and feel. In Quantum Entanglement, particles that have interacted with each other at some point (for example photons or electrons), always retain something of their original connection. No matter how great the distance or time between these particles, they always keep that connection as long as they are isolated. Also, the act of measurement of one particle in a system affects the state of another particle that may be far away. That means that things and people we come in contact with are always affected by us, whether they are 1,000 miles away or existing in a different universe. That is perhaps why it is difficult to end relationships, even bad ones! (This isn't implicit permission to continue negative relationships, but rather to help you recognize your feelings are legitimate, helping you to cope better while exiting. After all, you will form new and more positive connections on your path, and ideally these will be just as strong but more beneficial.)

The authors come from two different fields of study: science and metaphysics. Therefore we purposely examine multiple viewpoints in this literature. For the scientists among you, the preceding paragraphs may resonate strongly. Perhaps there is no real death because a version of ourselves continues to exist in a parallel universe. Crazier notions have been suggested in the past (electricity, television, you name it).

For those drawn to a faith-based or metaphysical framework, the notions of reincarnation, heaven, and spirit communication may feel closer to the truth. You might believe that a person's spirit leaves the physical body at death, moving into the spirit world and reuniting with loved ones, angels, and God.

Scientist or spiritualist… let's get started!

Time To Put The Phone Down

In today's society, constant distraction is altering the way we think. We expect immediate gratification, which is a root cause of feeling unsettled. Current thoughts on this problem are that we don't have any time to be bored…and boredom or relaxation often ignites invention and creativity. In order to relax we simply need to connect within and with the world around us. Meditation is a great way to do this, because it helps us disengage from the numerous devices surrounding us and re-engage with our innate ESP abilities.

Meditation is a practice of engaging and training the mind, or purposely striving to achieve a particular state of consciousness. Meditation will not necessarily tune you into an "ESP Frequency," so to speak, but it will allow you closer access and will also give you more control over your mental abilities by exercising your brain. Here are a few different types of meditation that you can practice:

1. Mindfulness
2. Zen
3. Transcendental Meditation
4. Kundalini Yoga
5. Qi Gong
6. Guided Visualization
7. Trance-based Practices
8. Heart Rhythm Meditation

9. Primordial Sound Meditation

Researchers are continuously uncovering additional benefits of meditation and its affects on the brain. The amygdala, the region associated with emotion and emotional memories, recovers more quickly from stress and trauma in a person who meditates. Studies show that gray matter (the area associated with self-awareness and passion) thickens with meditation, and the same study shows that regions associated with stress shrink. Meditation can also help to reset your "default mode" of being lost in thought or worry to a mode that is more fully active in the present moment. Countless studies show that meditation increases the ability to focus.

Do I Have To Sit In A Cave For Years Before I Can Use My ESP?

Nope! As you're starting to see, you are already a functioning psychic. You just may not know it yet if you don't understand how information from the sixth sense presents itself. Part of the learning process is investigating the roles of the conscious and subconscious mind. Once you know how to concentrate on your subconscious mind by connecting with the conscious-mind interface, you'll see for yourself that you're already tuned in. You can start to use meditation right away to tap into this always-available and valuable resource. Meditation familiarizes you with how subtle information "comes in," and it helps you develop your ability to concentrate on the information you access. In addition, the brainwave changes brought about by meditation help reshape your brain to be more attuned to this part of your mind.

The ability to recognize the differences between the conscious and subconscious minds is a key factor in developing your ESP. First, you need to get comfortable with the notion that metaphysics (and all it encompasses) makes sense in a different way than you're used to. It will make sense in the grand scheme of existence, but will likely seem out of place at the time you receive it. Second, understand that personal metaphysical experiences come through our sixth sense/intuition, which we know to be centered in the subconscious mind. Right alongside the subconscious mind is our imagination. The center for emotions, the limbic system also plays a role; the limbic center is very much involved in our instinctual functions and actions. That's why ESP can feel like your imagination, or that your instincts are pushing you one way while your logical mind is pushing you another way. Skeptics the world-over complain that metaphysical experiences can't be charted, measured, or reproduced—the scientific basis of experimentation. But in fact these experiences can be documented if you understand how to accept them for what they are and record them for analysis in the proper manner.

The conscious (thinking) mind does four things: rationalizes (come up with reasons for things that happen to us or that happen in the world), analyzes and solves problems (regardless of the scope or scale of the problem), powers our will (because we have to THINK about doing or not-doing something), and houses our working memory. Working memory, or short-term memory, is a very shallow "bucket" where we hold thoughts we need to regurgitate immediately, like a phone number we can't write down.

Hypnotists are taught that the conscious mind's job is to reject unfamiliar information. Automatically. Every time. There is a gateway between the conscious and subconscious mind called the Critical Factor. When information comes up against it, the Critical Factor asks, "Do I believe this? Is it familiar? Is it in my vernacular?" If yes, the Critical Factor door swings open and the information is allowed into the subconscious mind to have immediate effect on what is already stored there. If the answer is no, the information is immediately rejected. No matter how logical, provable, or evident, if the person doesn't believe what they're being told or shown, they'll reject it.

You may notice among your friends and colleagues that some people are "no" people and some people are "yes" people. "No" people have a tendency to shut down new or unfamiliar ideas, and you can imagine that their gateway between the conscious mind and subconscious mind is pretty thick and rigid. Saying yes or being open to contrary views and ideas in your conscious mind and daily life will help your conscious-subconscious connection flourish. Essentially, getting comfortable with cognitive dissonance is an exercise that can change your overall thinking.

The subconscious (instinctive, emotive) mind does everything else: it holds our beliefs, habits, subliminal instructions to the body, intuition, imagination, emotions, long-term memories, skills, etc. The conscious mind won't let anything new in here because that is not the conscious mind's job: its marching orders are to deal with what is happening now, and to bring information out from the subconscious mind when needed (where do I hang my keys, how do I get home, etc.). The psychoanalyst community's current consensus is that the subconscious mind doesn't subject information in it to analysis or rationalization, and therefore would not listen to the conscious mind anyway. But, this is very much up for debate. It is generally agreed that these two parts of the mind aren't mutually exclusive but don't communicate with each other very well, and that in daily life we function with a one-way street from our sub-conscious mind to our conscious mind. Our challenge in psychic practice is to get information through or around the conscious mind.

Some quantum physicists believe that consciousness stems from how our cells are able to express "brain states" effectively, that the conscious mind is perhaps a cellular snapshot of the subconscious mind at any given time. It isn't possible to take a similar snapshot of the conscious mind expressing in the subconscious mind, because brain states don't change in that manner.

It makes perfect sense that you may not yet be convinced of your psychic potential. The conscious mind wants proof, yet you may never be convinced to your conscious mind's satisfaction. Whatever floats up from your imagination, intuition, prayers, faith, "energy" is always going to be questioned somewhat by the rational mind: but with practice and repetition we can teach it to be more accepting, just as it accepts our base instincts and body functions on a daily basis.

Some students feel a big void because they are trying to make sense of something we do not have a lot of current language to describe. Accept that there is work to be done in order to understand the different ways psychic impressions present. That will help you to be a little more forgiving and a little less demanding of the subtler experiences you are already having.

We can liken this to food in a few ways. For example, people may believe with all of their being that they simply do not like Brussels sprouts…until they taste with the taste buds they have today (they change) the modern way it is prepared, instead of tasting with their 8-year old buds when their parents boiled the heck out of it until it was indeed disgusting. They don't actually accept it, and good luck trying to convince them! Of course if convinced, there is no changing the opinion back…do you see? It is also like tasting something new—the difference between wines for example. Or maybe learning to taste the subtle differences in other spices, if you're used to salt—which is clear, identifiable and unmistakable. Or try not using salt at all and concentrating on how a plain piece of steamed kale tastes different from a piece of steamed chard. It requires less "doing" and more "feeling." Feelings are almost always much subtler than actions. And the good news is that we can train ourselves to get more from a subtle experience if we back down from any existing rigid beliefs. At first when you back off salt everything tastes bland, all the same, boring. But over time, you may find that you begin to trust your sense of taste differently. Similarly, over time you can confidently identify what was previously feeling subtle or absent in your experience of ESP.

CHAPTER FOUR
Get Started—Who & What Am I?

First Things First

We get impressions from ESP through several channels, and knowing your strongest channel is the best way to start your practice. The following Visual / Auditory / Kinesthetic checklist will let you know the easiest natural channel for you.

Working as quickly as possible, and without over-analyzing your responses, check off all the phrases that apply to you. Go with your first impression.

Visual Channel
__ Like to keep written records
__ Typically read billboards while driving or riding
__ Put models together correctly using written directions
__ Follow written recipes easily when cooking
__ Review for a test by writing a summary
__ Write on napkins in a restaurant
__ Can put a bicycle together from written instructions
__ Commit zip code to memory by writing it
__ Use visual images to remember names
__ Am a bookworm
__ Plan the upcoming week by making a list
__ Prefer to get a map and find my own way in a strange city
__ Prefer reading/writing games like Scrabble

Auditory Channel
__ Prefer someone read instructions to you
__ Review for a test by reading notes aloud or by talking with others
__ Talk aloud when working out a math problem
__ Prefer listening to an audio book over reading the same material
__ Commit zip code to memory by saying it

___ Use rhyming words to remember names
___ Plan the upcoming week by talking it through with someone
___ Prefer oral instructions from an employer
___ Like to stop for directions in a strange city
___ Prefer talking/listening games
___ Keep up on news by listening to the radio
___ Able to concentrate deeply on what another person is saying
___ Use free time for talking with others

Kinesthetic Channel
___ Like to build things
___ Use sense of touch to put a model together
___ Can distinguish items by touch when blindfolded
___ Learned touch system rapidly in typing
___ Move with music
___ Doodle and draw on any available paper
___ Am an "outdoors" person
___ Move easily; am well coordinated
___ Like to feel texture of drapes and furniture in a room
___ Prefer movement games to games where one just sits
___ Find it fairly easy to keep fit physically
___ One of the fastest in a group to learn a physical skill
___ Use free time for physical activities

Now total up the number of checks under each heading:
_____ Total Visual
_____ Total Auditory
_____ Total Kinesthetic

If your highest number is in the Visual Channel column, you're predominantly clairvoyant. That means you'll get visual impressions most easily. As a beginner with ESP, you might want to practice the exercises (where possible) with your eyes closed or looking at a blank wall or some other neutral background.

If your highest number is in the Auditory Channel column, you're mostly clairaudient. Sounds may come across for you like an echo, or as if the source is far away.

If your highest number is in the Kinesthetic Channel column, you're mostly clairsentient. Awareness will just seem to pop into your head. You may feel a change in temperature in the room, or feel pressure on part of your body.

Most of us are some combination of these channels. Discover your strongest one, and then start to identify it and practice within its scope. Complete the sentence below, and say it to yourself often:

I am primarily CLAIR (voyant) (audient) (sentient).

It's important to remember that psychic impressions are extremely subtle. Don't wait to see something with the clarity you get when face to face with an object in the "real" world. You aren't likely to hear messages in the same way you hear a friend talk to you. And the impressions that come through clairsentience can be both subtle and strong, so it's important for you to remember that these are impressions of feelings.

In addition to understanding which "clair" will deliver sixth-sense impressions for you, psychic messages have several other characteristics you'll want to become familiar with. We'll delve into each of these more specifically a little later on, but for now take note:

- Psychic impressions (ESP) come in out of the blue. If you're primarily clairsentient, the thought will pop into your head even though you may be thinking about something entirely unrelated. For the clairvoyant, impressions will pop in visually. For the clairaudient, it might seem as though a sound suddenly reverberates.
- Psychic impressions are neutral. Just as other information from your five senses is neutral, so will be the sensations from your sixth sense. Your reaction may be positive or negative, but the data from your intuition will carry no emotional weight.
- As mentioned in the example above, psychic information will persist or nag at you, even though you try to dismiss it. Remember the neighbor who had the fire in his home from the beginning of the book? He tried to dismiss the notion of the fire, and even

when presented with an engineer's report and the assurances of the previous owner (who hadn't known about the fire covered up by the home's first owner), the feeling stayed with him.

• Your ESP may be accompanied by an urgent desire to take a certain action, or conversely, to avoid taking action. Your brain, DNA, and instinct are powerful and fight or flight plays a role here.

• You may not notice that you've responded to a psychic impression until after the fact.

CHAPTER FIVE
Mind & Body PushUps

Just like any skill, mastering ESP takes practice. One of the tools you'll be harnessing is your big, beautiful brain. You can increase your skill by strengthening your intuitive neural pathways, by challenging your brain as a whole, by improving your entire system with cardiovascular exercise, and by getting in tune with the universe (actually multiverse!) around you.

Actual Exercise? Hey, I Didn't Sign Up For This Part!
Read on... we may end up motivating you!
Let's first talk about brain exercises in general to clear up a few things. There's a great deal of disagreement in the scientific community about the effectiveness of brain exercises. Many studies show that brain exercises help increase your brain activity as a whole, often with long term, all-encompassing benefits. Other studies show that brain exercises only make you better at the one exercise (or "brain game") that you're repeating. And still other studies show no benefit at all. Basically, the scientific community hasn't figured this one out yet. We plan to continue playing "brain games" just in case the scientists in that camp are right! And, since it is widely agreed that training your brain to do one specific thing IS effective, in this chapter you'll find brain exercises that will help train specifically with ESP and communication (strengthening your intuitive neural pathways).

Our brain "wires" are neurons. Neurons are the building blocks of the brain, passing information through chemical and electrical signals. There are three parts of a neuron: dendrites, cell body, and an axon. Dendrites are the communication receiver, just like when you answer a call on your phone. The cell body is the processing center; think of the electronics within your phone doing all of the work. The axon is how information reaches the dendrites of other cells…like a long cable reaching out to another phone (we know, everything is cordless now but work with us!). When one part of the brain is working on something and needs to communicate with other parts of the brain or body, it sends nerve pulses that travel along neurons (like the signal a phone needs to receive in order to ring).

A neuron's favorite thing is myelin. Myelin is a fatty tissue that covers the axons. More myelin means the signals can travel faster and with more power. Myelin can be increased by consuming certain foods (i.e. whole grains, green tea, citrus, salmon, seaweed) and brain exercises (we'll help with these!).

We want to increase myelin on the right neurons, so it's best to find exercises that work for you, and for your natural channel (visual, auditory, or kinesthetic).

Now picture your friendly little neurons working away for you and passing along important information. What does this neuron network look like? Well, it's like miniature roads and tunnels running throughout your brain…AND these roads and tunnel routes can be changed to work better with ESP. More good news—rewiring, or changing these roads and tunnels is GREAT for your brain's longevity. You can also build "reserves" of neurons and connections by performing complex mental activity. Basically you can save up brain power for when you need it later!

Time for Brain Push-Ups

Here are a few exercises you can do at home to specifically target your cognitive intuition:

Pick a Card, Any Card…
This one can be done with a partner or alone. Look at two different cards. Then turn them to face away from you (or have your partner do this). Shuffle the cards and try to intentionally pick out one of the cards (for example if one card has a blue square and one card has a red circle, try to select the red circle card). Once you start to get good at this you can move to cards with more advanced pictures.

Guessing Amounts
This one works best with a partner. Place a certain number of items in a bowl or basket. Do not show your partner the contents. Ask your partner how many items there are, give some options (3, 7, 10, etc.).

Paper Shapes
This one works best with a partner. Draw a shape on a piece of paper, fold it up, and give your partner the piece of paper. (Tell them not to unfold the paper.) Show your partner different shapes, including the one you drew on the paper, and ask them to tell you what shape is on the piece of paper they are holding.

Specific Cards
You can buy (or make) HSP cards (HSP = Heightened Sensory Perception), which have various colors and/or shapes printed on them. Place a set of cards in a row face up (start with between 3 and 6). Then take matching cards, face down, and place them next to the card you believe is the same. Turn them over and see how you do!

What's Your Number, Baby?
This one definitely needs a partner. Sit facing your partner. In this exercise you have a giver and a receiver. The giver picks a number. The giver then needs to mentally picture the number going down a neuron, out to the end of a dendrite, and then instead of traveling over the synapse to another neuron inside their own brain, picture it going over the space between partners and onto the partner's neuron. The receiver takes a series of deep breaths, clearing the mind and when both partners feel they are ready, the receiver reports the number received.

You should exercise alone *and* in small groups. Set time aside each day to exercise yourself, and coordinate larger groups of people once a week. Find friends who are interested in the topic and exercise together, it will help your mirror neurons! Mirror neurons are a type of neuron that is activated the same way both when you act AND when you observe the same action. We truly reflect the world around us, therefore it's extremely important to surround yourself with good, like-minded people.

Brain Aerobics!

For *overall* brain health, you want to exercise your six main cognitive functions:

1. Memory
2. Attention
3. Language
4. Judgment and evaluation
5. Reasoning and computation
6. Problem solving and decision making

Exercising these abilities will help you keep a healthy, sharp brain for the full length of your life, and will also improve your overall abilities in each area, many of which are involved in ESP. For example, learning to pay attention is a significant way to help you improve your overall abilities. Start paying attention to your thoughts and intuition, as well as to incoming external signals. There are a number of resources online to help exercise these areas, just use the search term "brain games" to get started.

In addition, we recommend doing exercises to improve what is called the "useful field of view." The useful field of view is defined as the region of space over which a person can attend to information. It is also described as "the visual area over which information can be extracted at a brief glance without eye or head movements." How do you increase this functionality? You work to increase your "visual speed of processing," which will help you in many areas of your life. Many studies have been done with regard to driving ability, mostly aimed at increasing driver safety, particularly in the aging population.

Dr. Karlene Ball of the University of Birmingham, Alabama worked for three decades on studies focused on the useful field of view. She developed computerized training exercises that help in both driving and in other areas of daily functioning. In one study it was shown that "people 65 and older who are trained to process visual information more quickly are less likely to drive dangerously and are able to keep their licenses longer than a control group of older adults." Improving your useful field of view will increase your overall ability to attain visual information quickly from the physical world, but it will also increase your ability to process your psychic impressions.

Along the lines of this "quick glance" ability, recent research on numerosity and where it happens in the brain has opened up quite a conversation. Numerosity is the ability to map numbers in our brains and count items via a glance, without actually counting one-by-one. Why does this matter? The key has to do with brain mapping; if we can find where numerosity occurs, that means we can map high functions in the brain. This is exciting, because topological brain mapping may be able to map functions beyond our five senses, and make predictions on what lies ahead in scientific research for our sixth sense.

Sorry, You Aren't Going To Completely Get Out Of Exercising!

Although there is disagreement in the research community about whether or not brain exercises work, there is NO disagreement that physical exercise, good nutrition, mental stimulation and connection with others help brain function and longevity. Cardiovascular exercise in particular improves mood, memory, and helps avoid dementia down the road. Exercise creates activity in the brain that helps strengthen neurons and neural connections and increases overall brain health.

Scientists once thought that it was impossible to create new brain cells, but no more! Exercise results in the formation of new neurons—think of it like brand new baby neurons being born to help your own personal brain community.

Several studies have demonstrated that cardiovascular exercise improves brain functioning. For example, Arthur Kramer at the

University of Illinois, published research on how fitness affects brain development in children, showing that aerobic exercise increased the white matter in the brain—that's where your axons live! Healthier axons mean your psychic impressions have stronger roads to travel on.

Learning new things and being active has also been shown to improve brain health. The neurologist Dr. Joe Verghese published research after following almost 500 people for more than 20 years, observing their life activities and the relationship between their lifestyle and brain health. He found that by participating in mentally stimulating activities four times a week, such as dancing and interactive games, people had a 65-75% likelihood of remaining sharp (versus those that did not participate in these activities). Who doesn't want to be a sharper psychic?!

You Already ARE Psychic!

As you've read, you're already equipped with a fully-functioning sixth sense. It evolved with the human race and is a deeply embedded part of your awareness. So deeply embedded, that for many people it seems non-existent; but you wouldn't survive without your sixth sense. It's hard-wired into your brain.

After taking the Visual, Auditory, Kinesthetic (VAK) test, you now have a good idea of the channel which will most easily deliver intuitive information from your subconscious/non-rational/sixth sense mind into your conscious/rational mind. If you're mostly clairvoyant, you'll benefit by practicing the following exercises with your eyes closed, at least at the beginning. If you're clairaudient, practice in a setting without too much ambient noise so you'll be able to pick out the subtler "sounds" from your intuition. If you're clairsentient, make sure you're in a comfortable position in a comfortable environment so you can concentrate on the intuitive "feelings" you receive.

Don't worry too much about whether your specific channel is common to others. When I was developing my ESP, most of the books I read and teachers I studied with referred to "listening to the small, still voice within." "What do your guides say?" they would ask. "Listen for messages from your sixth sense," the books would advise. But I wasn't clairaudient! Not one bit! So for a very long time I thought I was failing because I couldn't "hear" what the authors and teachers were telling me I should be able to hear. It seemed I'd never be psychic. Then I came across the VAK test (the test you took in Chapter 4) and discovered my strongest channels were kinesthetic and visual. In fact, I didn't score even one check in the auditory column.

From that point on, whenever I read a book or took a class I just changed the words to suit my own best interpretation. As co-author of this book and being primarily clairsentient, I may refer to feelings or pictures more often than sounds. Please remember your own strongest channel; all the exercises here will enhance your sixth sense perception whether you're clairaudient, clairsentient, or clairvoyant.

***

Now you know the most fluent way your sixth sense will offer information to you. But how will you know *when* it's coming in? Let's refer to those examples we briefly outlined in the previous chapter. We'll take them one at a time:

Psychic impressions (ESP) come in out of the blue, apropos of nothing you've already been thinking about.

If you can trace your thought back to an earlier thought, you're engaged in simple linear thinking, or self-generated thought. Let's say I'm getting dressed for work and suddenly picture (hear, feel) myself being reprimanded by my boss. If I can backtrack that thought along a trail to an original mental statement, then it's a linear thought and not a psychic impression. For example: "Last time I came to the staff meeting in sneakers my boss made a snarky remark about looking professional ---> I can make the 9am meeting in my sneakers ----> I should probably wear sneakers and carry my shoes ---> I couldn't get a seat on that train and I had to stand the whole time ----> I'll be able to get the next train and still make it to work on time ----> That'll set me back ten minutes ---> I ought to iron this blouse." While getting dressed I made a mental note to iron my blouse, which led to subsequent thoughts about what would happen if I took the time to do so, ultimately bringing me to an idea that my boss would reprimand me. Psychic impressions present themselves differently. Using the same example, the idea of my boss reprimanding me would pop into my mind and be untraceable back to an original idea. Let's say I was washing the dishes and suddenly got the feeling (saw an image, heard a sound) that I was getting chewed out by the boss. If I can't explain what caused me to think that thought, I'm having a psychic impression. How is this helpful? Well, you can bet the minute I determine my intuition is warning me about a workplace reprimand, I'll leave the dishes in the sink and go take a look at my email, calendar, or even call a colleague to see if I've forgotten something important. If not, at the very least I'd be prepared when I got the phone call that I'd need to be defending or explaining myself to my supervisor.

Psychic impressions are neutral.

An impression may be followed by a slight feeling not a grandiose one. (Such as mild panic if we stick with the example above. "Uh oh! I'm forgetting something at work!") People can drive themselves crazy when they start thinking about ESP the way Hollywood shows it. In movies the poor unsuspecting character has an alarming mental experience, usually leading to horror, murder, or some other fearful event. The information that comes through the sixth sense, just like the information from your other five senses, is completely neutral. When your eyes detect a fire, you'll respond differently depending upon your state of mind and the circumstances. You'll feel joy and relief if you're lost in the freezing cold woods and you spot a campfire; frightened and panicked if your living room is on fire; happy and relaxed if it's candles on a birthday cake; cautious if you're watching children roasting marshmallows; or completely unfazed by it if someone else is tending an outdoor barbeque. The perception of the fire itself is neither good nor bad, but the myriad possible responses to it create a different experience every time. So how will you feel if your psychic eyes show you a fire? Your response also depends upon the circumstances of the vision. Imagine you're standing on a platform waiting to board a train, and suddenly (out of the blue), you have the feeling or impression of fire. You can't trace it back to an original thought, and your five senses aren't showing you evidence of fire. Inside you feel completely neutral, as though someone were just presenting you with the standalone concept of fire. In this case, we strongly suggest that you step back, let that train pass, and catch the next one! If however you're boarding the train and you start thinking about fire, and if your emotional response is one of completely freaking out so that you're asking yourself, "Am I having a psychic impression?" the answer is probably no, you're not.

Psychic information will persist, or nag at you, even though you try to dismiss it.

Even if you've managed to convince yourself that you're not having a psychic impression, and the thought keeps returning to you despite your attempts to dismiss it, you're probably perceiving something extrasensory. I remember one day looking at my calendar and making my usual confirmation calls for the next days' clients. I spoke with all four of them, concluding with a call to my 7pm client. The following day as I was preparing for my readings, I had the idea that I should call my 7pm client once again. I reminded myself that I'd called her just the day before and she had our appointment in her schedule. All throughout the day the thought would nag at me to call her again to confirm. Each time I talked myself out of it. I didn't want her to think I was absent-minded! Still, the thought kept popping into my head. When our 7pm appointment time began and she hadn't arrived, in a way I wasn't surprised. I make a living as a psychic after all! When I called her at 7:15 she was baffled to discover that she'd written our appointment down for the following day. I reminded her that we'd spoken just yesterday and she admitted that she wasn't really paying attention at the time because she was feeding her children when I'd called. Not really listening to me, she thought she was confirming an appointment for the next day.

Your ESP may be accompanied by an urgent desire to take a certain action or conversely, to avoid taking action.

Even though the information may come in as neutral, be prepared for the *impression* of urgency. Many years ago a man I know had been trying to sell his house with no luck. For months and months his house would be on the market without generating any interest from potential buyers. Finally, on the advice of his real estate agent he decided to take it off the market for a while. A few weeks later he was doing some yard work when he had the strong sense that he should put his house up for sale -- RIGHT NOW. He went into the garage, pulled out one of the old For Sale signs and pounded it into his lawn. Minutes later a car pulled up and to make a long story short, he sold his house to a couple passing by because he followed through with an urgent thought to do something. On the flip side of this, we can become aware of a strong sense NOT to do something. Every person I've worked with who has been the victim of a crime, accident or bad decision has said something along the lines of "I knew I shouldn't have gotten on that elevator/taken that boat ride/invested in that stock." The more we can learn to heed the impressions from our sixth sense the better equipped we will be to protect ourselves.

Psychic impressions may not be recognized until after the fact.

Because sixth sense information comes from the non-rational part of your mind, do not be surprised if it doesn't make sense right away. I recently did a reading for a business client of mine. She and her partner were considering a certain kind of financial partnership and I saw the impression of the Grinch who stole Christmas. I tried to clear my mind then to get a more reasonable psychic response to her question. But the Grinch wouldn't go away. At that point I knew that this was psychic information because it popped in, I couldn't dismiss it, and it was completely irrational! So I let the symbol evolve until I felt the meaning behind it, which was this: "This source is dressed up as something else. It's masquerading as something charitable and giving, but there is another agenda at play. If you begin to create a backup plan you won't lose anything." My client was confused by this, as was I. It didn't really make any sense to either one of us at the time. Two weeks later I got a call from her: the nonprofit status of the financial partner they were considering was coming under scrutiny from the IRS.

The rational mind tries very hard to make sense of irrational data. That is its job. When you engage your ESP, it is important to suspend judgment of your impressions. Remind yourself that they may not immediately make sense.

***

Now that you're armed with key identifying characteristics of ESP and you're aware of how your unique "clair" will offer you information, it's time to get started with some real-world practice.

I've always maintained with my clients and with the thousands of students I've had, that there are only four obstacles to ESP:

1. Not knowing what it feels like to be getting an impression
2. Inability to distinguish between thoughts and impressions
3. Lack of confidence
4. Inability to concentrate

We've just covered the first two obstacles. The following exercises are designed to wipe away the final two by giving you the opportunity to see how psychic you already are. Remember, you don't need to develop your sixth sense. This isn't a muscle that only grows when you stress it in the right way. Concentration, however, is something you can build with practice. The confidence-building exercises are designed specifically to show you that you are quite capable of getting and giving psychic impressions. The concentration exercises allow you to focus on your sixth sense impressions for longer periods.

Exercises to Build Confidence in Your Naturally Occurring ESP

We suggest you practice each exercise for a minimum of seven consecutive days, unless otherwise noted. Three weeks is optimal, but even trying these once will have an impact. Feel free to practice just a few or all of them; there's no need to learn one at a time.

Exercise #1: Truth or Lie?

It's always a good idea to recognize when your sixth sense is singing, "Liar, liar, pants on fire!" No lover will get away with cheating on you again. No unscrupulous mechanic will get away with charging "the little lady" for her made-up car problems. If you practice this exercise and trust the signals, no matter how subtle, no one will ever rip you off again. In addition to spotting lies and liars, you'll have the confidence to ask for that second date, leave the dog with a new dog-sitter, and choose the right college, career, or house for you.

Find several minutes when you won't be disturbed or pressured by the next task or appointment awaiting you. Take a deep breath or two, just to slow things down. Studies have shown that deep breathing introduces an alpha brainwave state, which is the meditative or relaxed mental state. Also, most of us tend to breathe rather shallowly most of the time. Taking deep breaths introduces more oxygen to the body and ultimately the brain, which helps us to feel more focused and alert. It might sound counterintuitive (no pun intended) to introduce a relaxed mental state and an alert state simultaneously, but remember that people who use their ESP fluently aren't sleepy, drowsy or lazy. They are just very alert to their sixth sense, and able to concentrate when needed on the information it provides.

Take two or three deep breaths to start, please.

If you're clairvoyant, it may help to close your eyes or gaze into the middle distance with soft focus, like daydreaming. We sometimes let our eyes relax while gazing down at the rug, at the woods outside the window, or just on a blank wall.

Call to mind something you *know* to be true. Your name, your heart's truest desire, the love you feel for your pets or children. Concentrate on that feeling by making it as full as you can. Dive right in. You'll begin to notice a certain feeling in your body (if you're clairsentient), or perhaps certain colors or shapes behind your closed eyelids (if you're clairvoyant), or a sound in your mind's ear (if you're clairaudient). Most people have a feeling sense about truth and falsehoods.

Continue to concentrate now on both the feeling you called up that represents truth to you and the feeling, picture, or sound that is accompanying it. Stay with it for as long as you can. You're successfully reinforcing a signal that you'll be able to rely on your whole life.

Open your eyes and take another two or three deep breaths, or stand up and stretch, take a walk around. The point is to create an end to the positive association portion. You can do this at two separate times during the day if that's more comfortable to you. At the very least, take a minute or two to do something different.

When you're ready, close your eyes and this time call to mind something you know to be *false*. Think of a time someone betrayed you, fooled you or lied to you; focus on the moment when you became aware of the lie. If you're in higher political office, or otherwise deeply involved in an organization or your community, you might recall an incident when you personally felt let down or lied to. If you don't have such an unfortunate incident in your life, think of a friend or family member who was lied to or betrayed. Uncomfortable as it may be, make this feeling as big as possible in your imagination now.

You'll soon notice another feeling, picture, or sound coming to you. One client hears a buzz in his ears whenever someone is fibbing. Another friend notices that her stomach seems to make a momentary clench. Yet another reports seeing small "tells" in the liar as if they were under a magnifying glass. This friend was once listening to her neighbor explain that he didn't know the law when he put in his fence, but all she could concentrate on was how he was rubbing his hands as if he were washing them, over and over. The gesture seemed huge to her clairvoyant perception, and she knew he was lying. Ordinarily my friend is non-confrontational and would rather accommodate than put up a stink. But being certain that he was taking advantage of her good nature gave her the backbone to insist that he move his fence off her property.

Pay close attention to whatever comes up for you. Focus on *both* the feeling of the lie as well as the response your sixth sense is offering you. This serves to create an association that will be invaluable in the future.

If you practice this exercise every day for seven days, it's perfectly okay to use different examples of the truth and lie. Both are universal and your sixth sense will recognize everything that falls into the each category.

It's very likely that you won't need to use this Truth/Lie check everyday, but we suggest that you work it into your daily life. The impression can be very subtle, so the more you use and experience it in all its levels of intensity, the more you will be able to rely on it confidently.

Practice This Skill in Everyday Life:

Practice your Yes and No feelings, clues or symbols everywhere you go. Try it in the grocery store when choosing between two different brands of something. Tune in to the subtle "good for me/not good for me" translation of the Truth/Lie check. You can also practice while you're watching TV or reading a book, especially mysteries; focus on a character and tune in to see if he is the villain. Practice at the mailbox before opening it. Tell yourself there are more bills than junk mail for example. Tune in to the "correct/incorrect" translation of the Truth/Lie check. Practice in any situation where you can check yourself right away.

How To Chart Your Progress:

When I first began training myself psychically, I had a special journal for my yes/no work. I noted what my yes/no signals were, and practiced everywhere. I began to see a pattern in my successes—I was almost 100% correct about pregnancies and the gender of babies. To this day clients seek me out when they are starting or expanding their families. Keeping a journal is a terrific way to chart your progress too, and to see what will emerge as your natural "specialty." For example, if you find that you are usually spot-on when it comes to spotting the bad guy in a book or movie, you may develop that skill towards helping the police solve crimes. You may see consistent success with numbers, animals, people, health issues, law or government policies. Don't worry too much about developing a weaker area—go with your strengths; the weaker areas may or may not catch up naturally. Personally, I have a flimsier ability to detect spirit phenomena in homes and buildings, so I refer those jobs to other psychics. Whenever anyone asks me about coming over to check out a haunting, I just say, "Sorry, I don't do houses!"

Exercise #2: Symbols

Regardless of your strongest "clair," most psychics get their information via symbols because symbols are a very efficient way to convey information. Without giving it too much thought, write down what these words (symbols) mean to you, with your very first impression. Your answers are translations of symbols from *your* library. There are no right or wrong answers in this word association exercise.

SYMBOL	YOUR MEANING
Moon	
Dog	
Shuttered Window	
Seedling	
A Cool Feeling on Skin	
Sound of Wheels Turning	
Fog Horn	
Tension in Stomach	

Below you'll see some examples from our workshops. To illustrate the various responses, we've placed more than one meaning in the YOUR MEANING column, separated by a semi-colon.

SYMBOL	YOUR MEANING
Moon	Mystery; Female; Turning tides
Dog	Friendship; Protection; Fear
Shuttered Window	Observe only; Illusion of safety
Seedling	Possibility; Nurturing
A Cool Feeling on Skin	Fresh start; Chilly reception
Sound of Wheels Turning	Think more about it; Travel
Fog Horn	Stay away; Guidance
Tension in Stomach	Financial Worries; False Information

It's interesting to note in these examples how varied peoples' libraries are. For one person, an impression of a dog gives them a feeling of friends, and for another person a dog symbolizes something to be feared. Everyone has had unique experiences, so the symbols can mean completely different things to different people. That's why you can't be wrong in your interpretation.

Practice This Skill In Everyday Life:

When you feel a strong emotion come up in your every day life, give it a symbol. If you feel a pain, have a headache or trouble sleeping, assign a separate symbol to each of those as well. When you associate your five-senses experiences with a symbol, your sixth-sense will use it during intuitive times. If you know someone who is pregnant, give that state a symbol. Anything that stands out for you, create an association with any symbol of your choosing. (You might want to start writing some of them down.) The point of this is not to rigidly adhere a sixth-sense impression with a written-in-stone symbol, but rather to practice communication between your subconscious and conscious minds, to limber up the translation from the deeper, less rational areas. Remember to consider your "clair" in this practice, too. If you're clairvoyant, let many of your symbols be visual; if clairaudient, let the symbols be predominantly sound.

How To Chart Your Progress:

It's a good idea to journal your symbol library, too, but don't get too rigid with this. Symbols will evolve and change as your confidence increases and you recognize subtleties in your impressions. It's a good idea to keep a piece of paper on hand while you're doing a reading so you can jot down the symbols you see. Afterwards, go over your doodles or notes and circle in a different color pen those that felt very clear to you, or those your client validated strongly. By singling them out in this fashion you are cementing their meaning while they are still fresh. You'll notice with each reading you do that the pages reflect greater numbers of different-colored circles, which indicates an expanding symbol library.

Exercise #3: The Psychic Box (Psychometry)

This one is so much fun, you'll want to practice all the time! It is a form of psychometry. (Use A. if you're practicing with a friend, B. if you're practicing alone)

3A: (Practicing With A Friend)

Ask a friend or family member to put an object into a small covered box while you're not looking. Hold the box in your left hand, clear your mind and speak the images, feelings, tastes, ideas, emotions, or scents that you're getting. Very often, especially in clairsentient impressions, an idea just seems to pop into your head. Voice it! Don't worry about trying to make sense of what you're getting at first. With more practice you can put all of that information into a sentence or message. For now, just say what you're picking up. You may ask your assistant to remain quiet until you're finished (and you'll feel when that is), or you may ask them to validate and validate only. Your friend may say "yes" to you in response only as a validation—you'll be amazed at how this seems to totally flood your extra-senses with more information. If the other person can't validate, ask them to say simply, "I don't know." It's very rare, by the way, for a psychic to tune into something the first time and get exact information. Psychics who help detectives offer clues: house numbers, names, feelings about what a victim might have experienced, but they very rarely deliver cut and dried information that's complete enough to close a case right there.

3B: (Practicing Alone)

Instead of using a box, gather several identical small paper lunch bags. Without dwelling too much on what you're doing, quickly walk around your home or room and put one small object in each bag. Direct your mind to something else for a while, or wait until tomorrow to continue. When you're ready, sit and lightly rest a bag in your hand without groping it to determine what's inside. Speak out loud into a tape recorder, or write down the feelings or images you're receiving. Don't spend a moment of this time trying to connect what you're saying with what you might remember having put in the bag. Let your mind flow with this exercise. You will know when you feel finished. Open the bag, and note the consistencies with the object (or its origin, use, or placement in your home) with what you've written or recorded.

Practice This Skill In Everyday Life:

I like to do this exercise at the mailbox. I put my left hand on it and briefly tune into what I will find there. Sometimes I get a feeling of a city or state, and see a return address reflecting that location. Sometimes I'll see the symbol of a skunk, which for me represents a newspaper. That symbol comes from a time when I was a little girl, and my aunt told me a joke that I had a hard time wrapping my mind around because the concept was a little abstract for my young mind. The joke was, "What's black and white and re(a)d all over? A newspaper!" Well, this didn't make any sense to me because I heard the word "read" as the color "red," probably because the colors "black" and "white" were also mentioned. In my mind I always saw a skunk, and when my aunt would deliver the punchline, I imagined a skunk reading the newspaper. Isn't it funny how symbols are created?

How To Chart Your Progress:

Information derived by psychometry, like all psychic impressions, is inexact. Accept that you may never be able define completely what it is that you are tuning in to. It is more important to note how subtle things are coming through to you more clearly as you develop your skill and confidence. For example, early on in your practice your partner puts a watch into the psychic box. If you sense a round shape and the feeling of a hand (even a human hand), that's pretty darn good. In the beginning, you'll be delivering the more obvious, stand-out characteristics. But if your partner puts that same watch back in the box after you've been practicing for a few months, your descriptions may be more nuanced. Perhaps you'll pick up on the feelings your partner has about that watch, or how happy he was to receive it on a special birthday. You might have a sense of it being broken, or that it belonged to someone who passed away. You'll know you're making progress when the information you get is more than just physical descriptors.

Exercise #4: Telepathy

Telepathy is a small part of your intuition, but a convenient and fun way to communicate. To begin taking advantage of your telepathic abilities (and we all have them, we just need to practice enough to recognize them), you need to have or create an open channel; simply setting the intention will do. An intentional message will be stronger than one sent unconsciously, so to do this one make sure both parties are ready and willing at the same time. If you're the receiver, be clear in your mind that you intend to receive the information that is to come.

Coordinate a 15-minute time slot with a friend or family member. They can be across the country or in the same room with you, as the psychic energy in us knows no time or space. Ask your friend to hold a picture from a magazine, or to draw an image on a new, blank piece of paper without showing it to you. When she's finished, have her imagine that she can send that picture to your mind. Any way she imagines this is perfectly fine; there is no right or wrong way. Clear your mind, center yourself, and intend that your own mind is open to receiving what is being sent. On your own clean sheet of paper, begin to draw shapes. Don't worry about making an actual image yet, just let your pencil move in the way your mind instructs. When you're finished, compare pictures. You'll find that you've drawn similar shapes, even if you haven't completely telepathized the entire composition.

To use telepathy to send a message to anyone, try this: Beginning with your eyes closed, bring the other person to mind using the five extra-senses. When the person occupies your awareness, imagine the message or image you wish to send. If it's a spoken message, repeat it a few times in your mind, strongly. Imagine them hearing or seeing it.

When someone doodles, the conscious mind is usually bypassed, clearing the way for information to come through. When you're doodling, it's likely that you're focusing on what someone was saying, or on something else entirely than what is taking place.

Try this exercise the next time you are on the phone with someone. Simply let yourself doodle and see what you come up with; you'll be amazed at the words, shapes and emotions your intuition expresses—sometimes revealing what the other person is really thinking, even if the words don't match. Remember that symbols, even those that appear while doodling, are for *you* to interpret.

Practice This Skill In Everyday Life:

This can be invaluable whether used with family, coworkers or friends. Concentrate hard on a person you need to hear from: see, hear, or think about her picking up the phone and calling you. Stand quietly next to a coworker or friend, and imagine plugging an invisible cord into them. Then try (subtly) to initiate a conversation on the topic that immediately comes to mind; you'll be so astonished (and delighted) by the consistent response, "I was just thinking about that!" or, "I was just going to ask you that!"

How To Chart Your Progress:

If you're on the receiving end of telepathic messages, it's hard to know until after the fact. Record your receptions as soon as you recognize them, as well those signals you send on purpose. Just like in the Yes/No exercise above, you'll notice stronger connections with a certain person, or at specific times of the day, or even regarding certain subjects. I seem to pick up on emotional highs in people, so I have a lot of fun calling my far-away sisters and saying, "Tell me the good news!" I always enjoy their reaction!

Exercise #5: Give Your Sixth Sense a Task

Your sixth sense has plenty of information just waiting to be accessed. Most of us just toss our lines in the water and wait to see what fish we catch, but those who are really adept at ESP also set their lines for specific fish.

Before falling asleep at night, pose a question to yourself—begin with one that you will know the outcome in a fairly short period of time. The answer may not come to you in a dream that you'll wake up and remember. It may come into your mind while you're showering, driving to work, or daydreaming in your cubicle. Be prepared for it to hit in an "Aha!" way, or through a symbolic form.

You can also do this while meditating, or simply by asking the question aloud and then going about your normal tasks.

Practice This Skill In Everyday Life:

Write your question down before asking, and when the answer comes to you note the form it came in and how long after you asked it.

How To Chart Your Progress:

You'll notice that the time between asking and answering gets shorter and shorter, and your interpretation of the answer gets easier and easier. Keep records in your journal, and above all, be sure to congratulate yourself robustly each time you make a connection. This serves to reinforce your own inner communication.

Exercise #6: The Concentration Game

Nothing is more important than being able to sustain focus on your sixth sense impressions, and that takes concentration. Not many of us have the ability to totally focus on something in this age of superfast technology, so this may seem boring or impossible. Keep at it anyway. Just like building any muscle, this will take a little bit of time, but practice means your powers of focus and concentration will grow by leaps and bounds.

Sit quietly and choose either a letter or a number. Close your eyes (if you're clairvoyant, or if you find it helpful) and try to hold that letter or number in front of you for as long as possible. Don't be surprised if you are only able to get one or two seconds out of it. Try to push it a little more each time you practice.

Practice This Skill Everyday Life:

Try focusing on what someone is saying without letting your mind wander. Really listen. Both of you will feel more connected and stronger. Try to give an email or work task 100% of your attention without letting distractions interfere. Do one thing at a time with as many activities as you can (but not at the same time!).

How To Chart Your Progress:

This is a subtle experience, but you'll notice your ability to focus on that number or letter gets longer and longer. Your understanding of conversations will improve, and stress will go down in direct proportion to your ability to concentrate. You'll stop procrastinating and you'll be much clearer on your own goals, and which steps to take to get there. You'll manage time better, and your relationships will improve, along with your intuitive abilities.

CHAPTER SIX
How Does This Fit in My Daily Life?

OCD For Breakfast, Lunch, and Dinner
    We have a zillion messages coming to us every minute in this society full of mobile devices and social media. How do we decide where to focus our attention? In addition to the messages coming through to us, distant radio stations and cellular towers are broadcasting hundreds of different radio waves and wireless frequencies (wifi) all around us. At any given instant, your office, car or living room is full of these radio waves. How to sift through all that is coming at you? You broadcast messages and receive external messages, how can you make the best use of it all? How can you listen to the right "channel" or the right parallel frequency? Learn to think about these things differently, modify your reactions, and practice.
    A radio can only broadcast one frequency at a time; these other frequencies are not in phase with each other. Each station has a different frequency, a different energy. As a result, you can only tune into one broadcast at a time on your radio. It is important to become aware of all of the available channels around you. Relating this to quantum physics for a moment, in our universe we are tuned into the frequency that corresponds to physical reality. But an infinite number of parallel realities coexist with us in the same room, there are an infinite number of universes at our fingertips, although we cannot easily tune into them. With practice, you will be able to sift through all of this available information more quickly, and by being open you will be able to "tune-in" to things you may not have known existed. It is a two-pronged approach—allow yourself to "hear" the variety of messages coming in to you, and then be able "choose" the right one for the situation. You have already begun to work on recognizing messages and impressions when they come to you.

Let's take this metaphor another step further. When you turn on your radio and choose a channel to listen to, your psychic and physical ears become attuned to the message and volume, and if you listen regularly enough, you also develop a set of expectations about what sort of information you're likely to receive.

Imagine that you've chosen to listen to a classical music station. You can reasonably predict that the music will fall into a certain style. Now imagine you've decided you will listen for the clarinets—suddenly you've become fascinated by them. In the beginning, you might not be able to tell a clarinet from an oboe, but over time, with training, persistence and concentration, you'll be able recognize and easily distinguish between the clarinet and other instruments. As you get more familiar with the sounds of the clarinet, you'll see how it adds information to the piece of music. In one song it will feel poignant and bittersweet; in another, agitated and alarming.

This ability to develop discernment is the reason the authors are so confident that with exercise (there's that word again) and practice, our readers will be able to determine which channel receives ESP signals strongest, how to recognize pertinent signals among all the other noise, and how the message is meant to inform us.

Learning how to recognize different incoming channels will change what is called the "pre-conscious" mind as well. The pre-conscious mind, a reasonably respected area of Sigmund Freud's research, is different from the conscious and the subconscious, although tied to both. This part of your mind helps to anticipate circumstances and aids reflex reactions. The pre-conscious deals with the environment (sights, sounds, external stimuli) and internal motivations (values, interests, goals) and how they produce a reaction. Yes, you will have more messages to wade through, but you'll get good at it and appreciate the extra available information. Learning to trust your pre-conscious mind is similar to trusting your instincts. Understanding the different aspects of your mind will give you more to work with, and will help grease the wheel of your brain as it becomes a psychic channel powerhouse.

Open the pathways in your brain to allow for more impressions and trust what comes to you. The more you do, the more the wiring in your brain will co-operate!

Tip: Make sure to include something creative in your training. For example, use art as a form of brainstorming to get your mind ready to think creatively. And get off your computers and your devices for a while. Research shows that all of these devices and their messaging are highly addictive. During brain scans, the same area of the brain that lights up for heroin in a heroin addict lights up when the subject receives a message on social media! So turn off your electronics and turn up your creativity.

Time For Brain Pilates!

Now that you have the basic building blocks, you can strengthen your abilities! You are really in the process of rewiring your brain, and you can help to hardwire some of the channels in your brain that are helpful for ESP. But we warn you, just like changing a habit, it may feel uncomfortable at first. Eventually the new way will become comfortable and the old way will feel uncomfortable.

Remember the comparison of the wiring of neurons in your brain to roads and tunnels? Well, now consider that the more you drive the road the more permanent the road becomes. And actually USING the road, traveling the road becomes easier, just as driving the same route all the time becomes automatic.

Don't worry, you won't do any damage to yourself by rewiring. The important parts that dictate day-to-day functions won't be changed—that type of activity is hard-wired. Our instincts are pretty much hard-wired too. What we can do is to learn to listen to and trust our instincts. In the Stone Age, instincts kept people alive. It was a very uncertain world and people were at the mercy of weather, natural disasters, and predators. Instinct was the number one thing humans relied on for survival. We have lost a lot of that by being surrounded by modern conveniences—let's get it back.

Life In Color

Trust what you "see" in people and understand the messages they are sending whether they are conscious, sub-conscious, or other-conscious. You have probably heard of auras, but you may not know that some people are born with the ability to see auras due to a scientific phenomenon called synesthesia.

Synesthesia is a neurological condition in which two or more bodily senses are coupled. For example, synesthetes can see or taste a sound, feel a taste, or associate people with a particular color.

To explain this, first let's cover some basics of how senses work in the brain. The brain has dedicated areas that are specialized for certain functions: One area for sound, one area for taste, one area for the identification of letters and numbers, and so on. Some research suggests that increased cross-talk between these different regions leads to synethesia. Increased "cross-talk" between different brain regions that are specialized for specific functions is usually innate, but can be developed.

There are several different types of synesthesia. Following are some examples:

1. Sound-to-Color Synesthesia
2. Grapheme Synesthesia
3. Number-Form Synesthesia
4. Personification
5. Lexical-Gustatory Synesthesia

To further elaborate let's examine some of the forms. In grapheme (color synesthesia), letters and/or numbers are seen with an inherent color. For a person with synesthesia, usually the color associated with a letter or number does not change. The letter or number they see or write is always a particular color. For example, if red is associated with the letter "E" then a word may look like this: Evergreen. In another form, ordinal linguistic personification, numbers, days of the week, or months of the year can evoke certain personalities. It follows that a color could be associated with certain perceived personalities if the area of the brain that processes color and the area of the brain that identifies personality "cross-talked," thus evoking a visible aura.

What does this do for you? Well, by practicing methods to achieve synesthesia, the limbic system can orchestrate a fusing of the senses for the right and left brain experiences. With continued practice, a synesthetic appreciation can eventually develop in you and give you one more tool with which to harness your innate creative psychic abilities. Increasing your control over the wiring of your brain gives you more control over functionality in general, and improves communication and reception of psychic impressions.

Developing Synesthesia Exercise #1:

Take the word IMPROVEMENT and think of it for a minute. Say it aloud and play with it. Use it in several phrases. Roll the word on your tongue. Is it stale, fresh or aromatic? How does it feel? Is it long, short, fat or skinny? Does it have a texture of smooth, soft, hard, pointed, mushy or prickly? How heavy is it? Does it evoke visual or auditory images?

Developing Synesthesia Exercise #2:

Assume a comfortable position and completely relax. Now ask someone to play a provocative piece of classical music that you're not already familiar with. As the music begins, open up all your senses to it. Imagine that your skin is hearing and feeling the texture of each note as the music flows over and through you. Imagine that your nose is smelling the flavor of it, and your mouth is drinking in each tone, and savoring the taste of it. Visualize an array of colors swirling around you in brilliant hues as the music is played. Use your powers of visualization, and let the music sweep through all of your senses. Allow a kinesthetic involvement, and move your hands and body if you wish. Write down what each piece or phrase smells, tastes or feels like in emotional way.

CHAPTER SEVEN
The Ethics of ESP

Time To Get Serious (Don't Worry, Not Too Serious)

Although we are not psychologists or doctors, we can take direction from the general principles of these disciplines, as our clients may have similar expectations. The APA (American Psychological Association) and the AMA (American Medical Association) have some appropriate guidelines to evaluate and discuss. We've broken down a few principles for your consideration below.

Beneficence and Nonmaleficence

Our scientific and professional judgments and actions may affect the lives of others; therefore a person counseling another person must strive not to do any harm to the subject. We must also take care that our own health and well being is intact in order to bring the most sound judgment to the table.

Fidelity and Responsibility

We must be aware of our professional and scientific responsibilities to society and to the communities we serve, accept responsibility for our behavior and avoid conflicts of interest. We must also be aware of our colleagues' actions, and safeguard each other. Volunteering our services is also a part of the big picture; always make sure to give back.

Integrity

We must strive for accuracy, honesty and truthfulness in our practice. Fraud or misrepresentation of information is unacceptable. Don't be afraid to say "I don't know." Nobody has all the answers, and "I don't know" is a valid answer.

Justice

It's simple, be fair and just. Treat everybody equally. Everyone has the right to improve their life and to commune with the universe.

Respect for People's Rights and Dignity

All people have the right to privacy and dignity. Autonomous decision-making is not available to people living within communities with extreme hardship, and this should be taken into consideration when fostering relationships. There are many cultural and role differences out there, and we must be mindful and thoughtful about these differences.

Choice

Always remember that it is okay to say "no" to people you are uncomfortable taking on as clients. The environment in which you engage your practice is also your choice.

Law

We must respect the law. And equally important: when a law is recognized to be harmful to a client or community of clients, steps should be taken to bring the nature of this harm to the attention of the appropriate legislature.

Take these principles with you in your daily life and practice. Not only is it the right thing to do, it will give you confidence.

***

I don't receive negative information, such as when or how a person is going to die. I receive information that may be perceived as negative only if I am also given redeeming circumstances, or how to avoid that outcome. How your client takes your information and makes any changes in their lifestyle or habits is up to them. If you do get a feeling of something unpleasant impending, ask your inner guidance how it can be avoided or remediated before giving the message to the client. If you see a car accident, ask for information about where and when, what the weather's like, what the other car looks like, etc. That way you can give your client enough information to be prepared, and at the very least, to buckle his seatbelt! *Never* give a message that leaves a client feeling disempowered. *Never* leave a client feeling hopeless or doomed. If that is your impression, keep asking for more so that you can deliver some mitigating information and turn the message into something more positive.

Another "Don't:" Never give advice. We are not in the business of telling people how to run their lives or what choices to make. It's best to give your client his or her answers in the form of conditional statements: "If you avoid the mid-town tunnel on Friday, you'll be escaping a fender-bender;" not: "Don't take the mid-town tunnel on Friday." Or, "If you change careers from landscaping to real estate, I see you relocating to Florida;" not "I think you should change careers because real estate in Florida looks like a better choice." Remind your clients (and yourself, if necessary) that intuition should add to good judgment, not replace it.

One thing I'm sure to ask my clients, especially if they ask me a question with potentially upsetting answers, is, "Are you sure you want to know the answer to that?" I tell them I don't seek out or receive negative information, but if they're asking me about a loved one or their own situation, I make sure to assess whether they really want to know.

If you are asked a question with the potential for an emotionally devastating outcome, be sure to clearly ask the client if they are prepared for the answer they don't want. Remind them that you are bound to share what you get, if they ask for it.

Finally, on the subject of relationships, be careful about how you express information regarding infidelity. If a client comes to you and asks if his partner is cheating on him, be sure to reframe the question. Ask, "Do you suspect your partner is cheating on you?" If he responds in the affirmative, and your reading gives you the same information, confirm his suspicions *only*. Here's an example:

CORRECT
Client: Is my wife cheating on me with a guy from work?
Psychic: Do you suspect she is?
Client: Yes.
Psychic (Sensing "yes"): I feel your suspicions are correct.

INCORRECT
Client: Is my wife cheating on me with a guy from work?
Psychic: (Sensing "yes"): Yes. She's with Bob from Purchasing.

Client goes home and confronts his wife, who then sues you for ruining her marriage. Client then confronts Bob and does him bodily harm, who then sues you for endangering his life.

This goes for any questions that might result in a legal matter or a situation where someone's reputation is threatened: "Did someone steal from me?" "Is someone lying to me?" Questions of this nature should always be reframed so that you affirm or deny the client's suspicions.

It's always a good idea to put a disclaimer on your website or fliers. Even in the middle of a reading, whether you're doing it for a friend or a paying client, reiterate that you are not ethically or legally allowed to give advice. Here is the one I use:

"Psychic readings are for entertainment only. All psychic readings are intended to offer insight into a person's personal life and do not in any way constitute legal, financial, medical or professional advice. By engaging in a reading with any psychic, you understand that psychic experts do not diagnose illnesses, including questions pertaining to death."

In New York State, fortune-telling is still a class B misdemeanor, punishable by 90 days in jail or a $500 fine. The law prohibits anyone from claiming to tell fortunes, exorcise curses, or manipulate occult powers except "as part of a show or exhibition solely for the purpose of entertainment or amusement." Fortune-telling fraud—a long and highly orchestrated con designed to sap huge sums of money from trusting victims, usually by telling them they have a curse placed on them—falls into a different category than fortune-telling alone, and it can lead to far more serious charges.

There is no licensing or certification for any sixth-sense work, so it's up to us to govern ourselves. NEVER tell anyone they have been cursed and that you can remove it. For that matter, don't believe anyone who tells you that YOU have been cursed! Quite often those who are seeking intuitive guidance are emotionally vulnerable, and the power of suggestion, no matter how casually given, might have a huge impact on the client. I remember hearing a story from one of my teachers many years ago. He told us about a woman in his neighborhood who read tea leaves. A young man who had recently lost his fiancee to another man was visiting New York from Ohio and went to the psychic hoping for some good news about his romantic future. Instead, she predicted his death within three months. The young man was so shocked that he went home and immediately put his affairs in order. Within three months he did die, though the autopsy revealed that there was nothing wrong with him. My teacher believed—and as a hypnotist who sees the impact suggestions can have, I believe too—this man had no other reason to perish other than that he had accepted a terrible suggestion which his mind and body carried out.

You have a responsibility to leave your clients or friends feeling *better*. I like to tell my clients that "forewarned is forearmed," and any information they get from me carries only as much weight as whatever other data they have collected, so they can make confident choices.

You can also protect yourself by recording all readings and keeping them for a period of time, and by limiting your clients' access to your services. I recommend that my clients let six months to one year pass before visiting me again. I have had to turn away a small number of clients because I could sense that they were surrendering their own responsibility and trying to transfer it to me. One young woman called me repeatedly about whether and how to spend her money, even after I reminded her that I was not legally permitted to advise her. A man I used to work with called me too frequently for help making business decisions.

Are You Ready To Do Battle?

It's important to stick to your guns and trust your instinct—after all, isn't that what we are imparting in our work? Practice what you preach *and* be 100% yourself.

"If your number one goal is to make sure that everyone likes and approves of you, then you risk sacrificing your uniqueness, and, therefore, your excellence." (Author unknown.)

It is not important to make every single person believe in what you do, no matter what part of your life you are considering. The path of least resistance is often the correct path. On this particular topic, the path of least resistance will lead you to the part of the population that needs you most—those who are open to psychic abilities, brain training, and the multiverse. So there is no need to try to convince people of things they don't want to hear.

If you come to the table comfortable in your own skin, you bring 100% of you. This allows you to be true to yourself, and allows people to really know you and all you have to offer. This will also help you say "no" when you need to. The fear of being disliked makes people say "yes" when they should not. By being true to yourself, you will be more comfortable exploring your feelings and your psychic skills. "Yes" and "no" messages will become clearer as you practice, and if you use them correctly in daily life.

Candor is a great thing to bring to your everyday life in general, and when you are exploring psychic abilities. You would be surprised how much your candor may help other people. That said, you should listen to others without judging; this will open you up to concentrate on the true message that a person is giving to you.

Also, set your expectations correctly and know that you will face harsh people and harsh comments. It is easy to provide compassion and give someone the gift of ESP when they are treating you with kindness and respect. The true test is to treat someone fairly because that is the right thing to do, not because you will get something in return. By following this rule, not only will you have the confidence that you are doing the right thing and helping to make your ESP communication more consistent, you will inspire others to follow suit.

***

"The most courageous act is still to think for yourself. Aloud."

Coco Chanel

It is important to know when to speak up. It is also important to know when to walk away. You have the power to turn down clients, and to walk away from those that are fiercely negative about your belief system. It's your world.

Be prepared for challenge when or if you announce what you're practicing. As we've discussed in earlier parts of this book, there are many Doubting Thomases for whom departure from accepted norms feels destabilizing and frightening. If you know your great-grandmother would be thoroughly freaked out if you told her you were seeing the dead, do the responsible thing and don't mention it. Dinner at your wife's boss's house is probably not a good time, either; save the shock-value announcements for another time. Be true to yourself, but remember that not everyone is as comfortable with ESP as you are becoming. For some, ESP goes completely against their religion. For others, the topic is interesting although they may be embarrassed to ask about it, or share their own experiences if they feel others are judging.

Consider presenting the sixth sense topic as one *you* are exploring, and don't even bother trying to convince someone who seems stridently against it. For those who are open to it, suggest books to read (like this one!) or suggest a private conversation. Remember that nobody knows *everything* about the sixth sense, so keep yourself open to learning from others, too.

We'll Go Out With A Bang

Let's expand a little on the multiverse theory here to leave you something more to think about in your psychic practice and exploration.

As mentioned before, modern cosmological thought allows for the existence of what is known as the multiverse. There are different interpretations of exactly what this might be. So, for our purposes we're going to define an individual universe as basically a subset of the multiverse, and we'll refer to our universe as the reality which we perceive and exist within (also a subset of the multiverse).

Many scientists maintain that other universes cannot be contacted from our universe; and, that to any inhabitants of these other universes, their universe would appear to be The Universe. But more and more scientists (as well as countless various professionals) are acknowledging that there may be a loophole in all of this.

In some models, these other universes are like bubbles that have ballooned off of another universe, like a side bubble forming from the wall of an inflated balloon, pinching off its source as it grows.

Another model is an amoeba splitting in two. In this case, it makes sense that the physical constants of that balloon universe would be the same as those of the parent universe from which it ballooned.

However, in other theories (such as parallel and many worlds versions), each universe could have different local physical constants, as the value of such constants could well be determined during the initial phase of inflation from whatever their source point (but not a bubble parent universe).

Many people think that the idea of the multiverse is similar to the 'many worlds' theory, in which an infinite number of universes exist, each one playing out an alternative version of what might have happened if a different decision had been taken. For example, if a coin toss landed one way rather than the other, or what happens if someone turns left instead of right, etc.

Reduced to it's absolute level, then, this type of multiverse would have endless (too many to digest!) opportunities for psychic impressions to be delivered, as there should be a huge number of alternative universes, not just for those decisions taken by sentient humans, but, in fact, for every particle reaction that has ever occurred since this universe was created. In essence, there would be a universe for every particle that decayed or did not decay, reacted/did not react, etc.—although they would not be called psychic impressions but rather communication resulting from intersections of universes.

This leads to the premise of an infinite number of alternative universes where anything and everything that can happen or will happen has already happened or will happen.

For example, if there is a separate and discrete universe for all possible events, like an infinite number of monkeys with an infinite number typewriters, where all possible events have been played out and are stored in some great cosmic memory bank...which version of you is You? Which version of me is Me? Which version of reality is Reality? Which version of the universe is the Universe? What if you are actually communicating with your self...is there another you in another universe simply giving you the information you are asking for (through psychic impressions)?

Perhaps, *you* are the memory of experiences that make up the being you believe to be You.

And perhaps these experiences are the sum of the choices you make and that trace your path across/through/between the infinite array of universes within the multiverse.

What if every single person's pathway through the multiverse is like a random walk, gathering the experiences of the choices that are made, irrespective of which universe those choices are made in? (For example, while we are sitting typing, we know/remember this because we are doing it in another universe we are not typing. But in this universe, we do not know/remember this because the choice we made was to sit and type.) When we type and send an email, the receiver has no knowledge/memory of us emailing them yet. At that point, our pathways cross over and travel along the same vector for some time (during which, perhaps, a billion more universes are created as a result of other people/particles/events occurring). These events in any of those universes, may have perhaps affected us, but did not affect us in our universe. So we continue on our critical pathway vector through the multiverse, unaffected by events that are not tangible to us, or rather, seem to have no bearing on our pathway.

As each event occurs that creates a new universe, the entire universe is recreated in the new alternative universe, so a copying/replication of all particles occurs. This leads to entanglement, which joins particles in one universe with those in another. This energy is picked up in some way...perhaps this leads to an interpretation of data not meant for the listener, perhaps perceived by some as voices in the darkness, snippets of falsely remembered things, seeing things that are echoes of another universe...but actually may be communication across the cosmos that we call psychic impressions.

Or perhaps such events occur when one person's pathway vector crosses another's unintentionally...not in tandem but more in the manner of a small boat passing over the unseen wake of a ship in the night, the interaction is real and is experienced, but known only to the one who crossed the ripple unseen.

And as the universe is simply the sum of the perceived realities of each individual person as they meander their way through the array of the multiverse, each person is, in effect, a pathway trail, crossing and meeting and sharing information. In essence, each of us is a separate energy construct, and energy can flow and carry data. From the stories of horses that won't enter certain barns where someone previously hanged themselves, to people who report seeing ghostly trails, to the feeling that a conversation has been had before, all are examples of pathways crossing and sharing some energy from one pathway to another. You, the reader, now know how to find your way through the universe to whatever you seek.

BOOM!!

ABOUT PRISCILLA KERESEY

Priscilla Keresey is recognized throughout the country as one of the most accurate, compassionate, and sought-after evidential mediums, connecting the physical and spiritual worlds for individuals and in group demonstrations. She considers herself a practical psychic, and enjoys helping her clients achieve their business and personal goals. Priscilla teaches workshops on developing psychic ability and offers training for mediums in her message circles.

Priscilla is certified by the National Guild of Hypnotists as an Advanced Clinical Hypnotist. She created and taught a highly-effective six-week program on personal empowerment for female inmates in the New York State Correctional System. Priscilla also teaches self-hypnosis to her private clients and in workshops, with a focus on building self-esteem as the first step to all positive change.

Ordained as an interfaith minister in Israel in 2005, Priscilla acts as guest minister at local Spiritualist churches. She is a keynote speaker at conferences on the topics of Reclaiming Your Connection to the Divine, Mind Power, and Creating Success & Prosperity.

Priscilla is the author of two books on her experiences as a medium. She also wrote the popular **Live & Learn Guides**™ series and is the author/producer of **Live & Learn Guides**™ self-hypnosis audio files (more information on the next pages). She resides in New York.

To contact Priscilla, use one of the following methods:

Online	Address
www.fixyourscreweduplife.com	Priscilla Keresey
www.apracticalpsychic.com	c/o Live & Learn
www.liveandlearnguides.com	P.O. Box 226
www.viahypnosis.com	Putnam Valley, NY 10579

ABOUT SHAE MONTGOMERY

Shae Montgomery knows that the most important thing you can do for the Multiverse is to be you. She has reached this conclusion through the university of life, obtaining a Master of Science, running part of a successful Silicone Valley technology company, putting people's faces back together as an Anaplastologist, and working in Hurricane disaster relief when she can.

Shae Montgomery, MS received her Bachelor of Arts degree from San Francisco State University in 2007. Her Master of Science degree was earned in 2009 from the University of Illinois at Chicago Graduate School of Biomedical Visualization where she pursued a dual Masters track in Anaplastology and Medical Animation. Her research included acquisition of 3D anatomical data in clinical settings. In animations and teaching she specializes in visually communicating complex medical processes, anatomical pathologies, and neural communication pathways.

Early on in her life, Originally from the Washington DC area, Shay Kilby traveled around the US and parts of Europe primarily solo, studying Photography, mingling with Art & Music scenes in towns like New Orleans and Chicago, and jumping between interests in art, science, and nature. Putting herself through school working 2 and sometimes 3 jobs at a time she gained her first degree – a BFA in Photography from San Francisco State University. She went on to gain Masters of Science as previously mentioned and she encourages you to find light in even the darkest of times, on the off chance it pays off and you get to stick around.

Her current areas of research are in 3D Printing and Astrophysics. She enjoys finding the endless connections between 3D imaging and prosthetics, earth-based science and astrophysics. She now resides in the greater Washington DC area and when she isn't working, she spends her time boating around the nearby Chesapeake Bay with her dogs, doing yoga on paddleboards, co-facilitating a Siblings of the Mentally ill group for NAMI and finding ways to incorporate the multiverse into everyday life.

ALSO BY PRISCILLA KERESEY

It Will All Make Sense When You're Dead
Messages From Our Loved Ones In The Spirit World

Psychic Medium Priscilla Keresey delivers stories, connections, and messages of hope and reunion in this light-hearted, personal book. After a brief tale of her own introduction to the paranormal, the author shares funny, poignant, and insightful words straight from the spirit people themselves. Together, the living and the dead seek forgiveness, solve family mysteries, find closure, settle scores, and come together for birthdays, anniversaries, and graduations. Quoting directly from her readings and séances, Priscilla reports the spirit perspective on mental illness, suicide, religion, and even the Afterlife itself. For those readers interested in developing their own spirit communication skills, the last section of the book offers meditations and exercises used by the author herself, both personally and with her students.

It Will All Make Sense When You're Dead is chock-full of simple and entertaining wisdom, showing us how to live for today, with light hearts and kindness.

Nobody Gets Out Of This Alive
More Messages From Our Loved Ones In The Spirit World

Psychic Medium Priscilla Keresey returns with more messages from our loved ones in Heaven. With her customary humor, the author shares insightful commentary from our deceased friends and family on the topics of staying healthy, making mistakes, resolving problems, what Heaven feels like, and more. In three distinct sections, The Human Condition, the Spirit Condition, and Frequently Asked Questions, Priscilla quotes directly from her readings and séances. Calling her work "translating for the dead," the author extends hope and wisdom from both human and animal deceased. Each section includes free pre-recorded meditations for physical, mental, and spiritual health and one to help you connect with your own loved ones in the Afterlife.

The messages in this book inspire us to embrace the notion that although our physical journeys will one day end, life continues, love survives, and all is forgiven. After all, as the spirit people remind us continually, **Nobody Gets Out Of This Alive!**

Fix Your Screwed-Up Life
Recover Your Inherent Self-Esteem and
Start Living the Life of Your Dreams

Are you frustrated with the state of your relationships, money, work, or just plain bad luck? Are you worried about the future, dogged by the past, or stymied in your present efforts? Worry no more, Dear Reader! In these pages you'll discover the one foundational problem that you didn't even know was holding you back, and choose from several plans to reverse it. You'll soon be able to pull out of that rut and get back on track no matter how impossible it seems right now. You'll once again wake up in the morning feeling happy, empowered, and excited about the possibilities the future holds.

Follow the simple steps in this book and you will experience better physical health, banish limiting beliefs and doubts, and reprogram yourself to create the career, relationship, health, and bank account that you want. If you're ready to start living the life of your dreams, keep reading to discover the simple secret to *Fix Your Screwed-Up Life*....

Ordering information can be found by clicking the "Products" link at **www.liveandlearnguides.com**

AND...

The Live & Learn Guides™ Series

Mapping Your Destiny:
How to Use the Amazing Power of Intention
Do you ever wonder why some people seem to have it all? Are you having trouble getting some traction on your dreams? The quickest, most efficient way to define and create the life you want is by setting intentions. In this book you'll learn:
- How to ask for what you want (set an intention)
- What not to do when you set goals or intentions
- The no-fail formula for designing the life you've always dreamed of

Getting the Money You Want, Not Just the Money You Need:
The Straight Path to Abundance
Have you read countless books on creating prosperity? Have you been to the seminars and done the visualizations and still find yourself struggling? Are you ready, now, to live in the relaxation and serenity that comes with knowing your financial needs are always taken care of? Then this is the last book you'll ever have to read! In this book you'll learn:
- What's been holding you back
- How to release it, once and for all
- The six steps to ongoing prosperity

Unveiling Your Psychic Powers:
Never-Before-Told Secrets of a Psychic Medium
You may not know this, but you're already psychic. Everyone is! While it's true that some individuals seem to have an effortless "gift" of seeing the future, reading the past, or communicating with spirits, every single one of us has the same tools and abilities. We just need to learn how to use them, practice, and develop confidence. In this book you'll learn how to:
- Distinguish a psychic impression from a self-generated thought
- Read other people and situations instantly and confidently
- Build confidence in your own abilities with fun, easy exercises

A Radically Successful You:
Easy New Ways to Achieve Any Goal, Fast!
Are you tired of watching your co-workers get promoted while you slog away day after day? Do you ever wonder how your friend manages to be in the right place at the right time, all the time? Do you compare yourself to self-starters or entrepreneurs and wonder what they have that you don't? You are about to learn the secrets of real success! You'll be one of the winners – guaranteed! In this book you'll learn:
- Why you may not be succeeding right now
- How successful people deal with obstacles
- Six steps to creating and programming automatic success habits

Your Total Health Solution:
3 Ways. 7 Days. Guaranteed!
Do you have trouble turning off the chatter and relaxing, either at your lunch break during the day or before bedtime at night? Does it feel like it's been years since you had energy to try something new or go after a long-held goal? Are you tired of feeling so rundown? If so, this book offers you ways to return to total health in a shorter time than you might have thought possible. In this book you'll learn:
- How to approach total health on three fronts simultaneously
- Easy-to-assimilate steps to take for physical health now
- What emotional health is, and how to begin experiencing it
- What mental balance is, and how you can find it
- What to do on Day 8 and beyond

Praying: Creating A Relationship With God
Did you ever wonder if prayer really works? Have you ever been concerned that you're not doing it "right?" Are you repeating by rote those old prayers from your childhood, wondering if it's really possible that God hears that droning? You're not alone. In this book you'll learn:
- What makes a prayer a prayer
- How to create a prayer, and how to pray
- How to know your prayers are being heard and answered
- The benefits of praying
- Whether traditional prayers or homemade prayers are right for you

The Live & Learn Guides™
Self-Hypnosis Audio Series

Change Your Mind, Change Your Fortune
Abundance begins in the mind. Creating a mindset that says "Yes!" to prosperity is the key first step in creating a fortune. Listen to this self-hypnosis audio file and prepare your mind to naturally, intuitively, and rapidly attract riches into your life.

Creating True Success
We get what we focus on, and when we focus on success we create more. Listening to this self-hypnosis audio file allows you to realize the many ways you're already succeeding, opening the door for increased opportunities to achieve your goals.

Lose Weight, Feel Great!
Everyone knows that a healthy weight is more than calories in and calories out. With this audio file you can re-program your inner mind and direct your body to the shape and weight you desire.

Relax & Rejuvenate
Relax, de-stress, and create total health in your body and mind. Listen, relax, and rejuvenate!

Meeting Your Spirit Guides
Take a serene and confident guided journey to the realm of the Higher Mind, where you'll connect with your spirit guides. This guided meditation is 100% successful in bringing you together with your team on the other side. Don't miss this opportunity to learn how your guides communicate with you!

Discovering Your Personal ESP
Everyone has the ability to tap into their sixth sense. With this guided meditation you'll discover what your ESP feels like, and you'll be able to practice several techniques to hone your inherent abilities and develop confidence.

Re-Train Your Brain For The Life You Deserve!
You can use your mind on purpose to create the life you deserve. Change your thinking and actually re-wire your brain to effortlessly direct positive changes in your life.

Live & Learn Guides™ are available in printed or electronic format. Live & Learn Guides™ self-hypnosis audio files are available in CD or MP3 format. Ordering information can be found by clicking the "Products" link at **www.liveandlearnguides.com**

www.ingramcontent.com/pod-product-compliance
Lightning Source LLC
Chambersburg PA
CBHW071409040426
42444CB00009B/2165